Galactic Secrets Unveiled:

The Discovery of Mysterious 'Old Smoker' Stars in a 10-Year Milky Way Study

Emmons Rivas

Copyright

Disclaimer

The content presented in this book is based on the latest available scientific research and observations. While every effort has been made to ensure accuracy and reliability, the field of astrophysics is dynamic, and new discoveries may have occurred since the book's completion.

Readers are encouraged to consider this work as a snapshot of our understanding of the universe at a specific point in time. Scientific knowledge evolves, and subsequent advancements or revisions may impact certain details presented in this book. It is recommended to consult reputable scientific sources for the latest information on topics discussed herein.

The author, Emmons Rivas, is an astrophysicist and science communicator, but this book is intended for general informational purposes. It does not replace professional advice or comprehensive scientific studies. The author and publisher are not responsible for any inaccuracies, omissions, or consequences arising from the use of information presented in this book.

Furthermore, views expressed in the book are those of the author and not necessarily reflective of the institutions or organizations associated with the 10-year galaxy study or the broader field of astrophysics.

Readers are advised to approach scientific knowledge with an open mind and to verify information through multiple sources for a comprehensive understanding of the ever-evolving cosmos.

About the Author

Emmons Rivas, a luminary in the realm of astrophysics and a celestial storyteller, is the guiding force behind this book. As both an accomplished astrophysicist and a captivating science communicator, Rivas merges the realms of scientific inquiry and cosmic wonder, inviting readers on an extraordinary journey through the cosmos.

Key Highlights

Pioneer in Astrophysics

Emmons Rivas has devoted their career to unraveling the enigmatic tapestry of the universe. Their pioneering contributions to the field of astrophysics, particularly in the study of our Milky Way galaxy, have left an indelible mark on the scientific community.

Dynamic Science Communicator

Beyond the laboratory, Rivas is a dynamic science communicator. Through a skillful blend of vivid storytelling and accessible language, they bridge the gap between complex astrophysical concepts and the curious minds of readers from all walks of life.

Beyond the Cosmos: A Personal Note from Emmons Rivas

In addition to their scientific pursuits, Emmons Rivas shares a personal note, inviting readers into the human side of the cosmic journey. From the thrill of discovery to the challenges of unraveling cosmic mysteries, Rivas provides a glimpse into the passion and curiosity that fuels their exploration of the unknown.

Table of Contents

Introduction

In the vast expanse of the cosmos, where stars silently orchestrate the symphony of the universe, astronomers embarked on a monumental journey—a 10-year galaxy study led by the University of Hertfordshire. Their mission: to unravel the mysteries concealed within almost a billion stars in the Milky Way, employing groundbreaking observational techniques and cutting-edge technology.

As the research unfolded, a tale emerged that defied conventional wisdom, introducing us to a new celestial entity—the "old smokers." These elderly giants, hidden at the heart of our galaxy, challenged our understanding of stellar behavior by remaining dormant for years, only to abruptly puff out colossal clouds of dust and gas into space.

This book, "Galactic Secrets Unveiled: The Discovery of Mysterious 'Old Smoker' Stars in a 10-Year Milky Way Study," delves into the intricacies of this extraordinary discovery. We will navigate through the collaborative efforts of a global team of scientists, explore the instrumental role of infrared light, and unravel the enigma surrounding these "old smokers" and their unexpected behavior.

As we embark on this cosmic journey, prepare to witness the unveiling of celestial phenomena that not only reshapes our comprehension of stars but also holds the potential to redefine the broader landscape of astrophysics. The narrative unfolds at the intersection of technology, curiosity, and the vastness of the cosmos, inviting you to join us on a voyage into the heart of stellar mysteries.

Overview of the 10-Year Galaxy Study

For a decade, a dedicated team of astronomers from around the globe undertook an ambitious mission—to scrutinize and understand the intricacies of nearly a billion stars within our Milky Way. Spearheaded by the University of Hertfordshire, this comprehensive study aimed to push the boundaries of astronomical knowledge, employing state-of-the-art observational techniques and advanced telescopic instruments.

The primary objective of the study was to shed light on the life cycles and behaviors of stars, with a particular emphasis on uncovering rarely seen phenomena. As the astronomers gazed into the night sky, they sought not only to identify newborn stars, known as protostars, undergoing stellar growth spurts but also to explore the hidden corners of the galaxy where enigmatic entities like the "old smokers" lurked.

The use of infrared light proved instrumental in this endeavor, allowing the team to pierce through the cosmic dust and gas that often obscures visibility in visible light observations. The Visible and Infrared Survey Telescope (VISTA), stationed in the Chilean Andes, became the lens through which this cosmic exploration unfolded.

This book chronicles the journey of the 10-year galaxy study, from its conceptualization to the intricate details of its execution. It unravels the challenges faced by the researchers, the technological innovations that fueled their quest, and the unexpected revelations that emerged from scrutinizing the heart of our galaxy. Through this exploration, we aim to provide a glimpse into the profound mysteries that the cosmos unfolded during this decade-long astronomical odyssey.

Importance of Infrared Light Observation

In the pursuit of understanding the cosmos, astronomers have harnessed the power of a different spectrum of light—Infrared. This shift in observational wavelength has proven to be a game-changer in unraveling the mysteries of celestial bodies, especially during the 10-year galaxy study led by the University of Hertfordshire.

Penetrating Cosmic Dust and Gas
Unlike visible light, which can be obstructed by dense clouds of cosmic dust and gas, infrared light possesses the unique ability to penetrate

these obscuring elements. This characteristic is invaluable in studying regions of the galaxy that would otherwise remain hidden.

Revealing Hidden Phenomena

Infrared observations have the capacity to unveil celestial events and entities that might go unnoticed in visible light. The study's focus on identifying protostars and the unexpected discovery of "old smokers" highlights the importance of this alternative wavelength in revealing the hidden secrets of the universe.

Observing Cool and Obscured Objects

Many astronomical objects, including cooler stars and regions shrouded in dust and gas, emit predominantly in the infrared spectrum. Infrared observations allow astronomers to study these cooler, often overlooked, components, providing a more comprehensive understanding of the stellar population within the galaxy.

Enhanced Precision in Stellar Analysis

Infrared light facilitates a more precise analysis of stellar spectra, offering insights into the composition, temperature, and other critical characteristics of celestial bodies. This precision is crucial for unraveling the complexities of stars, especially those exhibiting unique behaviors like the "old smokers."

Expanding the Astronomical Toolbox
Infrared observations complement traditional visible light studies, providing a more comprehensive view of the universe. The inclusion of infrared data enhances the astronomical toolbox, allowing researchers to paint a richer picture of cosmic phenomena and refine existing models of stellar evolution.

As we delve into the implications of the 10-year galaxy study, the crucial role played by infrared light becomes evident. It opens a new frontier in our exploration of the cosmos, offering a lens through which previously obscured celestial marvels come into focus, enriching our understanding of the universe's vast and diverse tapestry.

The Research Team

At the forefront of the 10-year galaxy study stood a diverse and dedicated group of astronomers, collectively working to unravel the celestial secrets embedded within the Milky Way. Led by Professor Philip Lucas of the University of Hertfordshire, this international team brought together minds from the United Kingdom, Chile, South Korea, Brazil, Germany, and Italy, forming a collaborative effort that spanned continents.

Global Collaboration

The study's success hinged on the collaborative spirit of astronomers from around the world. Drawing expertise from various nations enriched the project, ensuring a diverse range of perspectives and skill sets were brought to the table.

University of Hertfordshire Leadership

Professor Philip Lucas, the visionary leader of the operation, played a pivotal role in orchestrating the study. His expertise and leadership guided the team through the intricacies of the project, from conceptualization to execution.

Cutting-Edge Technology Experts

The team comprised individuals well-versed in the latest astronomical technologies. Their proficiency in handling advanced telescopes, such as the Visible and Infrared Survey Telescope (VISTA) in the Chilean Andes, was instrumental in the success of the observational aspect of the study.

Infrared Observations and Spectroscopy Specialists

With the emphasis on infrared light observations, the team included specialists in infrared astronomy. These experts played a crucial role in maximizing the potential of VISTA and extracting valuable insights from the infrared data.

Spectroscopic Analysis Team

As the study delved into individual stars and their spectra, a dedicated team focused on spectroscopic analysis. Utilizing the European

Southern Observatory's Very Large Telescope, they examined the unique signatures in the starlight, unraveling details about the composition, temperature, and behavior of the observed stars.

Cross-Disciplinary Collaboration

Astronomers with diverse backgrounds, ranging from protostar formation to stellar evolution, collaborated seamlessly. This cross-disciplinary approach fostered a holistic understanding of the celestial phenomena encountered during the study.

Contributions from Early Career Researchers

The research team welcomed the contributions of early career researchers, ensuring a blend of experienced astronomers and fresh perspectives. This dynamic mix enhanced creativity and innovation within the project.

Collaboration Across Borders

In the quest to explore the cosmos, the 10-year galaxy study exemplified the power of collaboration transcending geographical boundaries. Astronomers from the United Kingdom, Chile, South Korea, Brazil, Germany, and Italy converged their expertise, forming a dynamic international team that collectively sought to unravel the mysteries concealed within the Milky Way.

Diversity of Expertise

The collaboration brought together a diverse array of expertise, encompassing various aspects of astrophysics. This diversity ensured a comprehensive approach to the study, incorporating insights from different branches of astronomy and enriching the project with multifaceted perspectives.

Shared Resources and Infrastructure

Collaborating across borders enabled the utilization of shared resources and state-of-the-art infrastructure. The European Southern Observatory's Very Large Telescope and the Visible and Infrared Survey Telescope (VISTA) in the Chilean Andes, both integral to the study, exemplified the collective utilization of cutting-edge astronomical facilities.

Continuous Communication and Coordination

Effective collaboration necessitates seamless communication and coordination. The team employed advanced communication technologies to bridge time zones and continents, ensuring real-time collaboration despite the physical distance between team members.

Cross-Cultural Collaboration

Beyond professional expertise, the collaboration fostered cross-cultural interactions. Astronomers with diverse cultural backgrounds contributed unique perspectives, fostering an environment of mutual learning and understanding.

Overcoming Time and Logistical Challenges

Coordinating a decade-long study across borders presented time and logistical challenges. The team's ability to overcome these hurdles showcased the resilience and dedication of the collaborative effort.

Global Contribution to Scientific Knowledge

The collaboration resulted in a global contribution to scientific knowledge, with each participating nation leaving its imprint on the understanding of celestial phenomena. The shared discoveries not only advanced the field of astrophysics but also reinforced the importance of international cooperation in scientific endeavors.

Capacity Building and Future Collaborations

Collaborative projects of this scale contribute to capacity building within the global scientific community. The knowledge and skills exchanged during the study lay the foundation for future collaborative endeavors, ensuring a sustained impact on astronomical research.

Tools and Telescopes Used in the Study

The success of the 10-year galaxy study hinged not only on the expertise of the astronomers but also on the advanced tools and telescopes employed to explore the depths of the Milky Way. The combination of cutting-edge instruments and state-of-the-art observatories played a pivotal role in unraveling the celestial mysteries concealed within our galaxy.

Visible and Infrared Survey Telescope (VISTA)

The VISTA telescope, stationed in the Chilean Andes at the Cerro Paranal Observatory, emerged as a cornerstone of the study. Equipped with powerful infrared capabilities, VISTA provided a unique lens through which astronomers peered into regions of the Milky Way obscured by dust and gas, revealing the previously unseen.

European Southern Observatory's Very Large Telescope (VLT)

The VLT, situated in Chile's Atacama Desert, contributed to the study's success by enabling detailed spectroscopic analysis. Its ability to capture high-resolution spectra of individual stars played a crucial role in understanding the composition, temperature, and behavior of the observed celestial bodies.

Infrared Observational Techniques

Infrared light became the key to unlocking hidden cosmic phenomena. The team utilized advanced infrared observational techniques to penetrate cosmic dust and gas, allowing for the detection of cooler stars, protostars, and the unprecedented discovery of "old smokers."

Spectroscopy Tools

Spectroscopy, a fundamental technique in astrophysics, was employed extensively in the study. Spectroscopic tools, both on VISTA and VLT, provided detailed information about the chemical composition and physical properties of stars, offering insights into their evolutionary stages.

Global Positioning Systems (GPS) and Time-Keeping Instruments

Coordination across international borders demanded precise timekeeping. GPS systems and accurate time-keeping instruments were crucial for synchronizing observations, ensuring that the collaborative effort unfolded seamlessly over the decade-long duration of the study.

Data Analysis Software

The massive volume of data collected from the telescopes required sophisticated analysis. Advanced data analysis software allowed astronomers to sift through terabytes of information, identify patterns, and extract meaningful insights, contributing to the robustness of the study's findings.

Communication Technologies

Effective collaboration across borders relied on advanced communication technologies. Video conferencing, secure data sharing platforms, and real-time communication tools facilitated constant coordination among team members spread across different continents.

Computational Resources

High-performance computing resources were essential for processing and analyzing the extensive datasets generated by the telescopes. Computational capabilities played a vital role in modeling stellar behaviors and conducting simulations to complement observational findings.

Stellar Growth Spurts: Protostars Revealed

In the celestial dance of star formation, the emergence of protostars represents a pivotal and enigmatic phase. The 10-year galaxy study, led by the University of Hertfordshire, cast a spotlight on these celestial infants, capturing unprecedented insights into their growth spurts and transformative processes.

The Quest for Protostars

The study embarked on a mission to identify and understand the elusive protostars—newborn celestial entities undergoing significant growth. These protostars, often shrouded in dust and gas, pose challenges for conventional observational methods. The utilization of infrared light, particularly through the Visible and Infrared Survey Telescope (VISTA), enabled astronomers to pierce through cosmic obstructions and unveil these cosmic newborns.

Stellar Growth in Technicolor

Infrared observations provided a unique perspective, allowing astronomers to witness the vibrant colors associated with stellar growth. Protostars undergoing intense growth spurts exhibited a kaleidoscope of hues, indicating variations in temperature, density, and chemical composition within their forming structures.

Extreme Outbursts and Dynamic Transformations

The study uncovered dozens of protostars engaged in extreme outbursts over varying timescales—ranging from months to decades. These outbursts, akin to stellar tantrums, played a crucial role in the formation of new solar systems. The dynamic transformations observed during these events provided valuable clues about the intricate interplay of matter within the protostellar environments.

Role of Infrared Light

In the pursuit of protostars, the decision to focus on infrared light proved instrumental. The ability of infrared observations to penetrate the dusty veils surrounding these celestial nurseries allowed astronomers to witness protostars during their most active and transformative phases.

Spectroscopic Analysis of Protostars

The European Southern Observatory's Very Large Telescope (VLT) played a crucial role in dissecting the spectra of individual protostars. Spectroscopic analysis unveiled the chemical signatures within these stellar nurseries, shedding light on the composition of the gas and dust clouds giving birth to these cosmic entities.

Unprecedented Numbers of Protostars

The study's extensive reach and observational prowess resulted in the identification of a staggering number of protostars—far surpassing previous discoveries. The ability to capture the largest batch of erupting protostars to date allowed astronomers to analyze these phenomena throughout their evolutionary stages.

Challenges and Future Investigations

While the study revealed a wealth of information about protostars, it also posed new questions and challenges. The dynamic nature of these celestial objects and the mechanisms triggering their extreme outbursts remain topics ripe for future investigations, promising further revelations about the early stages of stellar evolution.

Identifying and Understanding Protostars

In the cosmic ballet of star formation, protostars emerge as protagonists, heralding the birth of new celestial entities. The 10-year galaxy study led by the University of Hertfordshire undertook a pioneering exploration to identify and comprehend these elusive cosmic infants, unraveling the intricacies of their formation.

Hidden Gems in Cosmic Clouds

Protostars are often concealed within dense clouds of gas and dust, making their identification a formidable challenge. The study employed innovative observational techniques, primarily leveraging the power of infrared light through the Visible and Infrared Survey Telescope (VISTA). In doing so, astronomers pierced through the cosmic veils, revealing protostars previously shrouded in obscurity.

Infrared Penetration and Protostar Revelation

The decision to focus on infrared light proved pivotal in this endeavor. Unlike visible light, infrared wavelengths can penetrate the dusty environments where protostars gestate. By utilizing the capabilities of VISTA, astronomers were able to unveil the presence of these celestial nurseries and witness protostars during their crucial formative stages.

Distinctive Signatures

Protostars exhibit distinctive signatures that set them apart from other stellar entities. Their infrared emissions, characterized by variations in temperature and luminosity, provide unique fingerprints. The study involved meticulous analysis of these signatures, allowing astronomers to differentiate and classify these celestial infants within the vast cosmic landscape.

Kaleidoscope of Colors

Infrared observations painted a vivid tapestry of colors associated with protostars undergoing intense growth spurts. The kaleidoscope of hues reflected variations in temperature, density, and chemical composition within the protostellar environments. This visual richness offered valuable insights into the dynamic processes shaping these nascent celestial bodies.

Spectroscopic Insights

Complementing infrared observations, the European Southern Observatory's Very Large Telescope (VLT) played a crucial role in dissecting the spectra of individual protostars. Spectroscopic analysis unveiled the chemical compositions of the surrounding gas and dust clouds, providing a deeper understanding of the conditions conducive to protostar formation.

Dynamic Outbursts and Evolutionary Clues

The study identified protostars engaged in extreme outbursts, offering a glimpse into their dynamic nature and evolutionary pathways. These outbursts, observed over varying timescales, provided clues about the processes influencing the growth and transformation of protostars as they evolve into mature stars.

Challenges and Future Prospects

While the study marked significant strides in identifying and understanding protostars, it also posed new challenges and questions. The dynamic and complex nature of these celestial entities prompts further investigations, beckoning astronomers to delve deeper into the mechanisms orchestrating the early stages of stellar evolution.

Unprecedented Eruptions of Protostars

In the cosmic theater of star formation, the 10-year galaxy study led by the University of Hertfordshire unveiled a spectacle of cosmic proportions—unprecedented eruptions of protostars. This extraordinary revelation provided a front-row seat to the dynamic and dramatic outbursts occurring during the formative stages of these celestial infants.

Astronomical Tantrums Unveiled

Protostars, typically associated with steady growth and evolution, astonished astronomers with their unexpected and extreme outbursts. The study identified a sizable number of protostars engaged in eruptions that exceeded the scale of any previous observations. These celestial tantrums, akin to stellar fireworks, became a central focus of the research.

Diverse Timescales of Outbursts

What set these eruptions apart was not only their scale but also the diverse timescales over which they unfolded. Some protostars exhibited rapid and intense outbursts over the course of months, while others underwent more prolonged eruptions lasting years or even decades. This variability added a layer of complexity to the understanding of protostar dynamics.

Infrared Revelation of Eruptive Events

Infrared observations, facilitated by the Visible and Infrared Survey Telescope (VISTA), played a pivotal role in capturing these eruptive events. The ability to peer through cosmic dust and gas enabled astronomers to witness the infrared signatures associated with the energetic processes occurring within protostars during these extreme outbursts.

Largest Batch of Erupting Protostars

The scale of the study allowed for the identification of the largest batch of erupting protostars to date. This comprehensive survey provided a unique opportunity to analyze the eruptions throughout their evolution— from the initial quiescent state, through the peak of brightness, and into the declining stage. Such a wealth of observational data offered invaluable insights into the full lifecycle of these eruptive phenomena.

Solar System-Sized Clouds of Dust

The eruptions manifested in the form of solar system-sized clouds of dust expelled into space. These cosmic belches, observed for the first time with such magnitude, presented a new dimension to our understanding of protostar behavior. The study raised questions about the mechanisms triggering these eruptions and the impact on the surrounding protostellar environment.

Significance for Stellar Evolution

The unprecedented eruptions of protostars hold profound significance for the broader understanding of stellar evolution. They shed light on the processes influencing the growth, mass accretion, and environmental impact of protostars, contributing essential pieces to the intricate puzzle of cosmic birth and development.

Challenges and Future Investigations

While the study brought forth a wealth of information, the nature and origin of these eruptions pose new challenges for astronomers. Understanding the underlying mechanisms and factors driving such extreme events remains a tantalizing avenue for future investigations, promising continued revelations in the field of astrophysics.

Infrared Insights: VISTA Telescope Unveiling Cosmic Mysteries

Amidst the vast cosmic expanse, the Visible and Infrared Survey Telescope for Astronomy (VISTA) emerged as a transformative instrument during the 10-year galaxy study led by the University of Hertfordshire. This cutting-edge telescope, nestled in the Chilean Andes at the Cerro Paranal Observatory, played a pivotal role in unraveling cosmic mysteries by providing unprecedented infrared insights.

Piercing Through Cosmic Veils

One of VISTA's distinctive capabilities lies in its proficiency in capturing infrared light. Unlike visible light, infrared wavelengths have the unique ability to penetrate cosmic dust and gas that often shroud celestial objects. This capacity allowed VISTA to unveil hidden phenomena, including protostars and elderly giants, that remained obscured in traditional observations.

Revelation of Protostars

VISTA's infrared observations were instrumental in identifying and characterizing protostars—celestial infants undergoing intense growth spurts. The telescope's ability to peer through the dense clouds surrounding these nascent stars provided astronomers with a clearer view of their formative stages, enriching our understanding of the early processes of stellar birth.

Unprecedented Eruptions Captured

The telescope's infrared prowess played a central role in capturing the unprecedented eruptions of protostars. By detecting the infrared signatures associated with these eruptive events, VISTA provided a comprehensive view of the dynamics unfolding during these cosmic tantrums. This unique insight into the lifecycle of erupting protostars marked a significant contribution to astrophysical knowledge.

Identification of "Old Smokers"

VISTA's keen infrared eyes were also instrumental in the identification of a new type of elderly giant star, affectionately nicknamed "old smokers." These stars, residing near the center of the Milky Way, had eluded previous observations. VISTA's ability to pierce through the dense regions of the galaxy allowed astronomers to spot these dim and enigmatic celestial entities.

Enhancing Spectroscopic Analyses

VISTA's contributions extended beyond direct observations. The telescope facilitated enhanced spectroscopic analyses, offering detailed insights into the chemical composition, temperature, and behavior of celestial objects. This capability became particularly valuable in the study's collaboration with the European Southern Observatory's Very Large Telescope (VLT) for in-depth spectroscopic examinations.

Decade-Long Sky Survey

VISTA's role in the decade-long sky survey, known as 'VISTA Variables in the Via Lactea' (VVV), underscored its endurance and reliability. This extensive survey covered nearly a billion stars, demonstrating the telescope's capacity for sustained observations and data collection over an extended period.

Impact on Galactic Understanding

The infrared insights gleaned from VISTA significantly impacted our understanding of the Milky Way's composition, dynamics, and stellar populations. By revealing hidden stars and providing detailed information about various celestial phenomena, VISTA enriched the tapestry of our galactic knowledge.

Role of Infrared Light in Studying Stars

In the expansive realm of astrophysics, the utilization of different wavelengths of light is akin to employing various tools to unravel the mysteries of the cosmos. The 10-year galaxy study, led by the University of Hertfordshire, underscored the pivotal role of infrared light in studying stars, offering a unique perspective that transcends the limitations of visible light observations.

Penetrating Cosmic Obstructions

One of the fundamental challenges in studying stars lies in the presence of cosmic obstructions such as dust and gas clouds. Visible light observations often encounter barriers, rendering certain celestial objects hidden or indistinct. Infrared light, with its longer wavelengths, possesses the ability to penetrate these cosmic veils, unveiling celestial phenomena that remain obscured in visible light observations.

Revelation of Hidden Protostars

Protostars, in their formative stages, are enveloped in dense clouds of gas and dust. Infrared observations, as demonstrated in the study through the Visible and Infrared Survey Telescope (VISTA), allow astronomers to peer through these obscuring layers. The revelation of hidden protostars, undergoing intense growth spurts, showcases the unique power of infrared light in exposing celestial nurseries.

Eruptions in Infrared Spectrum

The eruptions of protostars, captured in unprecedented detail during the study, emit distinct signatures in the infrared spectrum. Infrared observations facilitated the identification and analysis of these eruptions, offering insights into the energetic processes occurring within these celestial infants. This capability becomes particularly crucial for understanding the dynamic phases of stellar evolution.

Identification of Elderly Giants

The study's discovery of a new type of elderly giant stars, colloquially named "old smokers," near the galactic center was made possible through infrared observations. These elderly giants, which remained dim and enigmatic in visible light, were brought into focus by the penetrating nature of infrared light, emphasizing its role in uncovering the diverse population of stars within the Milky Way.

Enhanced Spectroscopic Analyses

Infrared light enriches spectroscopic analyses, providing detailed information about the chemical composition, temperature, and behavior of stars. The collaboration with the European Southern Observatory's Very Large

Telescope (VLT) exemplifies the synergy between infrared and spectroscopic observations, enhancing our understanding of stellar characteristics.

Extended Sky Survey Reach

The extended reach of infrared observations allows for comprehensive sky surveys covering large portions of the Milky Way. The 'VISTA Variables in the Via Lactea' (VVV) survey, conducted over a decade, leveraged infrared light to observe nearly a billion stars. This extensive survey broadens our knowledge of stellar populations and their distribution across the galaxy.

Insights into Galactic Dynamics

Infrared observations contribute significantly to our understanding of the dynamics within the Milky Way. By revealing hidden stars and capturing phenomena not discernible in visible light, infrared light becomes an invaluable tool for unraveling the complex interplay of stars, gas, and dust in our galactic neighborhood.

Visible and Infrared Survey Telescope (VISTA) in the Chilean Andes: Unveiling Cosmic Secrets

Nestled atop the majestic Cerro Paranal Observatory in the Chilean Andes, the Visible and Infrared Survey Telescope for Astronomy (VISTA) stands as a technological marvel, transcending the boundaries of conventional observation. The integral role played by VISTA in the 10-year galaxy study, led by the University of Hertfordshire, highlights its significance in unraveling the cosmic tapestry.

Strategic Location at Cerro Paranal

Perched at an altitude of 2,635 meters on Cerro Paranal, VISTA benefits from the pristine skies and minimal atmospheric interference prevalent in the Chilean Andes. This strategic location ensures optimal conditions for astronomical observations, enabling the telescope to capture clear and detailed images of celestial phenomena.

Infrared Capabilities

VISTA's primary strength lies in its advanced infrared capabilities. Equipped with a large near-infrared camera, the telescope operates in the infrared spectrum, allowing it to peer through cosmic obstacles such as dust and gas clouds. This unique attribute played a pivotal role in the study, enabling astronomers to unveil hidden celestial objects and phenomena.

Visible and Infrared Survey Telescope

True to its name, VISTA is designed to conduct both visible and infrared surveys of the night sky. While visible light observations provide essential information, the telescope's ability to switch seamlessly to infrared observations significantly enhances its versatility. This dual capability is crucial for studying a broad range of celestial objects, from distant galaxies to protostars within our Milky Way.

Contributions to the VVV Survey

VISTA played a central role in the 'VISTA Variables in the Via Lactea' (VVV) survey, a decade-long effort to observe nearly a billion stars across the Milky Way. The telescope's infrared prowess allowed astronomers to identify protostars, eruptive events, and hidden giants, contributing to a comprehensive understanding of stellar populations within our galaxy.

Eruptive Protostars and Old Smokers

The telescope's infrared observations were instrumental in capturing unprecedented eruptions of protostars, shedding light on their dynamic and transformative processes. Additionally, VISTA's capabilities led to the serendipitous discovery of a new type of elderly giant stars, affectionately named "old smokers," near the galactic center. This discovery showcased the telescope's efficacy in revealing enigmatic celestial objects.

Complementary Role with VLT

Collaborating seamlessly with the European Southern Observatory's Very Large Telescope (VLT), located nearby at Paranal Observatory, VISTA enriched the study through enhanced spectroscopic analyses. This collaboration allowed astronomers to delve deeper into the chemical composition and physical properties of observed celestial entities.

Decade-Long Dedication

VISTA's decade-long commitment to the study exemplifies its reliability and endurance. The telescope's sustained observations and data collection over an extended period contributed to the robustness of the research findings, reaffirming its status as a cornerstone instrument for long-term astronomical endeavors.

Unexpected Discovery: "Old Smokers"

In the vast cosmic theater, where stars silently weave their celestial narratives, the 10-year galaxy study led by the University of Hertfordshire delivered an unexpected twist with the discovery of a mysterious group of celestial giants—affectionately named "old smokers." This unforeseen revelation, occurring near the heart of the Milky Way, added a new and intriguing chapter to our understanding of stellar evolution.

The Enigmatic "Old Smokers"

Amidst the extensive observations of nearly a billion stars in the Milky Way, astronomers stumbled upon a cohort of red giant stars displaying perplexing behavior. These elderly giants, clustered near the galactic center, earned the moniker "old smokers" due to their tendency to sit dormant for years or even decades before suddenly puffing out colossal clouds of dust and gas into space.

Surprising Dimness and Invisibility

One of the distinctive characteristics of these "old smokers" was their surprising dimness and, at times, near invisibility. Their dormant phases left astronomers in the dark, and it was only when these celestial giants decided to exhale massive clouds of smoke that they became noticeable. This unexpected behavior challenged preconceptions about the predictability of stellar activities.

Solar System-Sized Cosmic Puffs

The eruptions of "old smokers" manifested in the form of solar system-sized clouds of dust and gas. These cosmic puffs, observed for the first time with such magnitude, raised intriguing questions about the mechanisms triggering these eruptions and the subsequent impact on the surrounding cosmic environment. The discovery added a layer of complexity to the understanding of late-stage stellar dynamics.

Concentration in the Nuclear Disc

A further nuance to the discovery lay in the concentration of these enigmatic "old smokers" in the innermost part of the Milky Way, known as the Nuclear Disc. This region, rich in heavy elements, became the stage for the cosmic performances of these elderly giants. The proximity to metal-rich environments hinted at potential connections between their behavior and the distribution of heavy elements in space.

Challenges to Existing Models

The existence of "old smokers" introduced a fresh challenge to existing stellar models. While conventional wisdom pointed to well-understood stellar types, the unexpected nature of these elderly giants suggested that stellar dynamics, particularly in later life stages, might be more diverse and intricate than previously thought.

Significance for Element Distribution

Beyond the captivating nature of the discovery itself, the "old smokers" held broader significance for our understanding of element distribution in space. Matter expelled from aging stars plays a vital role in seeding the next generation of stars and planets. The revelation that a new type of star contributes to this process hinted at potential implications for the spread of heavy elements in the galactic neighborhood.

Unraveling the Mysteries

As astronomers delve into the intricacies of "old smokers" and their cosmic performances, the unexpected discovery prompts a reevaluation of stellar evolution and the factors influencing late-stage behaviors. Unraveling the mysteries behind these elderly giants promises not only a deeper understanding of their individual dynamics but also potential insights into broader processes shaping our Milky Way galaxy.

Initial Observations and Goals of the 10-Year Galaxy Study

Embarking on a celestial odyssey, the 10-year galaxy study led by the University of Hertfordshire commenced with a set of initial observations and ambitious goals, seeking to unravel the secrets hidden within the vast tapestry of the Milky Way. The journey began with a vision, driven by the desire to push the boundaries of astronomical understanding and unveil cosmic phenomena that had eluded scrutiny.

Skyward Gaze and a Billion Stars

The study's initiation involved casting a metaphorical gaze skyward towards the rich expanse of the Milky Way. Armed with advanced observational tools and a spirit of scientific inquiry, astronomers set out to study nearly a billion stars scattered across our galactic neighborhood. The sheer scale of this endeavor reflected the commitment to comprehensively explore the diverse stellar population within the Milky Way.

Inauguration of the VVV Survey

A cornerstone of the study was the implementation of the 'VISTA Variables in the Via Lactea' (VVV) survey, a decade-long effort leveraging the capabilities of the Visible and Infrared Survey Telescope (VISTA). This survey aimed to conduct a meticulous examination of the variability in brightness among the surveyed stars, unraveling dynamic processes and transient events occurring within our galaxy.

Identification of Protostars

One of the primary goals was to identify and understand protostars—celestial infants in the early stages of formation. The study focused on utilizing infrared light, particularly through VISTA, to peer through the veils of cosmic dust and unveil protostars undergoing intense growth spurts. This quest sought to enrich our comprehension of the intricate processes governing stellar birth.

Pursuit of Eruptive Events

With a keen interest in dynamic celestial events, the study set out to capture the unprecedented eruptions of protostars. Observations aimed to track these eruptive events over varying timescales, offering a holistic view of the life cycle of these cosmic infants—from quiescent states to peak brightness and eventual decline. This pursuit promised insights into the energetic processes shaping protostellar evolution.

Exploration of Stellar Variability

Beyond protostars, the study sought to explore the broader realm of stellar variability. By meticulously analyzing changes in brightness among the surveyed stars, astronomers aimed to classify and understand various events, from well-understood phenomena to enigmatic occurrences. This exploration provided a comprehensive view of the dynamic nature of stars within the Milky Way.

Collaboration with VLT for Spectroscopic Analyses

Complementing the infrared observations of VISTA, the study engaged in a collaborative effort with the European Southern Observatory's Very Large Telescope (VLT). This collaboration aimed to conduct detailed spectroscopic analyses of selected celestial objects, unraveling the chemical composition and physical properties of stars, protostars, and eruptive events.

Unveiling Hidden Giants and Unexpected Discoveries

While the primary focus was on protostars and dynamic phenomena, the study remained open to the unforeseen. The discovery of a new type of elderly giant stars, playfully dubbed "old smokers," highlighted the serendipitous nature of astronomical exploration. This unexpected revelation underscored the importance of staying receptive to the cosmic surprises hidden within the vastness of space.

Description of "Old Smokers" and Their Peculiar Cosmic Ballet

In the celestial ballet of the Milky Way, a newfound group of celestial performers, aptly named "old smokers," took center stage with their enigmatic and peculiar behavior. These elderly giants, discovered during the 10-year galaxy study led by the University of Hertfordshire, exhibited a cosmic dance marked by periods of dormancy followed by sudden, awe-inspiring eruptions.

Dimness and Invisibility

"Old smokers" earned their name by virtue of their distinctive characteristic—sitting dormant and almost invisible for years or even decades. During these extended periods, these celestial elders kept a low cosmic profile, their dimness often rendering them nearly imperceptible in the vastness of the Milky Way.

Solar System-Sized Eruptions

The defining feature of "old smokers" lay in their propensity for abrupt and colossal eruptions. When these elderly giants decided to exhale, they did so in a grandiose fashion, puffing out clouds of dust and gas on a scale comparable to our entire solar system. This unexpected behavior defied conventional expectations and introduced a new dimension to late-stage stellar dynamics.

Unpredictable Cosmic Puffs

The eruptions of "old smokers" were akin to cosmic puffs, each event releasing substantial amounts of material into space. These solar system-sized clouds, once expelled, added an element of unpredictability to the otherwise serene existence of these aging stars. The origin and triggers of these cosmic puffs remained shrouded in mystery, adding to the intrigue surrounding these celestial entities.

Elderly Giants in the Galactic Center

"Old smokers" chose a specific locale for their cosmic performances—the innermost part of the Milky Way, known as the Nuclear Disc. Concentrated in this metal-rich region, these elderly giants defied expectations by showcasing their peculiar behavior in a region where stars tend to be richer in heavy elements. The concentration of "old smokers" in this unique environment posed additional questions about the interplay between stellar dynamics and element distribution.

Dynamics and Convection Currents

The mechanisms triggering the eruptions of "old smokers" remained a cosmic puzzle. Hypotheses suggested that convection currents and instabilities within these aging giants might be responsible for the sudden release of enormous columns of smoke. These solar system-sized clouds were speculated to be the result of localized disturbances on the surfaces of these elderly stars.

Extended Dim and Red Phases

Adding to the mystique, "old smokers" underwent extended phases of dimness and redness. Their quiet existence, often dim and red for several years, preceded the dramatic eruptions. This characteristic behavior made the detection and classification of these enigmatic stars a challenging task, further emphasizing their unexpected nature.

Cosmic Surprises and Stellar Diversity

The discovery of "old smokers" highlighted the rich diversity within the stellar population of the Milky Way. Their unexpected behavior challenged existing models of stellar evolution, underscoring the need for a more nuanced understanding of the late stages of stellar life. These celestial surprises offered a glimpse into the complex and varied ways in which stars, even in their later years, contribute to the cosmic symphony.

Spectroscopic Analysis with the Very Large Telescope (VLT): Unveiling Cosmic Secrets

In the realm of astronomical exploration, the quest to understand the intricacies of celestial objects often involves more than mere observation. The 10-year galaxy study, spearheaded by the University of Hertfordshire, harnessed the power of spectroscopic analysis, enriching its findings through a collaborative venture with the European Southern Observatory's Very Large Telescope (VLT). This partnership brought forth a deeper understanding of the chemical composition, physical properties, and dynamic processes shaping stars and their cosmic companions.

Spectral Insights into Celestial Objects

Spectroscopy serves as a powerful tool in astronomy, allowing scientists to dissect the light emitted or absorbed by celestial objects. The VLT, perched atop the Paranal Observatory in Chile, played a pivotal role in obtaining high-resolution spectra, enabling astronomers to unravel the unique signatures embedded in the light from stars, protostars, and eruptive events.

Complementary Collaboration with VISTA

The collaborative synergy between VLT and VISTA, the Visible and Infrared Survey Telescope, was instrumental in the 10-year galaxy study. While VISTA excelled in infrared observations, VLT complemented these insights with detailed spectroscopic analyses. This combination provided a comprehensive view of celestial phenomena, from the visible to the infrared spectrum.

Fine-Tuning Understanding of Protostars

Protostars, undergoing dynamic growth spurts, were subjected to meticulous scrutiny through spectroscopic analysis. VLT's ability to capture detailed spectra facilitated the fine-tuning of our understanding of protostellar processes. Chemical composition, temperature variations, and other vital parameters were discerned, contributing to a more nuanced portrait of these celestial infants.

Eruptive Events in High-Resolution Detail

Eruptive events, captured in unprecedented detail during the study, underwent thorough examination through VLT's spectroscopic capabilities. The high-resolution spectra obtained allowed astronomers to delve into the intricacies of these cosmic outbursts. Insights into the composition of ejected materials and the energetic processes involved were gleaned, enhancing our comprehension of stellar dynamics.

Characterizing the "Old Smokers"

The enigmatic "old smokers," discovered near the galactic center, were subjected to detailed spectroscopic analysis. VLT's precise measurements facilitated the characterization of these elderly giants, offering insights into their chemical makeup and shedding light on the mechanisms triggering their peculiar behavior. The spectroscopic data became a key to unraveling the mysteries surrounding these unexpected stellar entities.

Exploration of Stellar Variability

Stellar variability, a central theme in the study, benefited from the spectroscopic prowess of VLT. The varying brightness observed among stars was coupled with detailed spectral information, allowing astronomers to categorize and understand the diverse range of events occurring within the Milky Way. This exploration expanded our knowledge of the dynamic nature of stars.

Contributions to Galactic Dynamics

Beyond individual objects, VLT's spectroscopic analyses contributed to our understanding of galactic dynamics. By examining the chemical composition and physical properties of stars across the Milky Way, astronomers gained insights into the broader processes shaping the evolution of our galactic neighborhood.

Unveiling Cosmic Mysteries: Utilizing the Very Large Telescope (VLT) at the European Southern Observatory

In the pursuit of unraveling the secrets of the cosmos, the 10-year galaxy study led by the University of Hertfordshire harnessed the unparalleled capabilities of the European Southern Observatory's Very Large Telescope (VLT). Perched atop the Paranal Observatory in the Chilean Andes, the VLT emerged as a celestial sentinel, offering astronomers a window into the intricate details of stars, protostars, and eruptive events across the Milky Way.

Optical Excellence at Paranal

Situated at the high-altitude Paranal Observatory, the VLT stands as a testament to optical excellence. Comprising four individual telescopes with 8.2-meter mirrors, the VLT operates as an interconnected array, providing astronomers with unparalleled resolution and sensitivity. This optical prowess proved invaluable in capturing the nuances of celestial phenomena during the decade-long study.

Interdisciplinary Collaboration with VISTA

Collaborating seamlessly with the Visible and Infrared Survey Telescope (VISTA), the VLT brought its optical acumen to complement VISTA's infrared observations. This interdisciplinary approach allowed astronomers to explore cosmic events across a broad spectrum, from visible light to infrared, enriching the study's findings with a comprehensive understanding of the observed celestial objects.

Spectroscopic Precision

The VLT's spectroscopic capabilities played a pivotal role in the study, offering a detailed look at the chemical composition, temperature variations, and physical properties of celestial entities. High-resolution spectroscopy facilitated the fine-tuning of observations, enabling astronomers to decipher the unique spectral signatures imprinted in the light emitted or absorbed by stars and other cosmic phenomena.

Capturing Protostellar Birth Pangs

As the study delved into the realms of stellar nurseries and protostars, the VLT's precision allowed astronomers to capture the birth pangs of these celestial infants. The telescope's ability to obtain detailed spectra contributed to a nuanced understanding of protostellar processes, shedding light on the mechanisms governing their early growth and development.

Unprecedented Views of Eruptive Events

Eruptive events, characterized by sudden and intense outbursts, were brought into unprecedented focus through the lens of the VLT. The telescope's high-resolution imaging and spectroscopy provided a detailed chronicle of these cosmic explosions, offering insights into the composition of ejected materials and the energetic dynamics at play during such events.

Spectral Exploration of "Old Smokers"

The unexpected discovery of "old smokers," the elderly giants exhibiting peculiar behavior, became a focal point for VLT's spectroscopic prowess. The telescope's precise measurements contributed to the spectral exploration of these enigmatic stars, aiding in characterizing their chemical makeup and unraveling the mysteries behind their dormant phases and sudden eruptions.

Contribution to Galactic Understanding

Beyond individual objects, the VLT's contributions extended to the broader understanding of galactic dynamics. By scrutinizing the chemical composition and physical properties of stars scattered across the Milky Way, astronomers gained insights into the evolutionary processes shaping our galactic neighborhood, enriching the study's comprehensive exploration.

Understanding Stellar Spectra and Light Wavelengths

In the intricate dance of the cosmos, stars communicate their stories through the language of light. Understanding stellar spectra and the nuances of light wavelengths is akin to deciphering the celestial script written across the vast expanse of the universe. The 10-year galaxy study, led by the University of Hertfordshire, delved into this cosmic language, leveraging advanced tools to unravel the mysteries concealed within the intricate patterns of stellar light.

Celestial Fingerprints

Stellar spectra are akin to fingerprints, unique signatures imprinted in the light emitted or absorbed by stars. These spectral fingerprints carry vital information about a star's chemical composition, temperature, density, and motion. Deciphering these cosmic fingerprints unveils the individual characteristics of each star, contributing to our understanding of stellar diversity.

Continuous and Absorption Spectra

Stellar spectra manifest in two primary forms: continuous and absorption spectra. Continuous spectra result from the thermal radiation emitted by stars, creating a seamless rainbow of colors. In contrast, absorption spectra occur when elements in a star's atmosphere absorb specific wavelengths, leaving dark lines or "absorption lines" that serve as celestial clues for astronomers.

The Prism of Light Wavelengths

Light, the messenger of celestial information, encompasses a spectrum of wavelengths. From the short wavelengths of ultraviolet light to the longer wavelengths of infrared light, stars emit radiation across this vast spectrum. Analyzing this prism of light allows astronomers to explore the diverse processes unfolding within stars, from nuclear fusion in their cores to the dynamic interactions within their atmospheres.

Spectroscopy Unveils Composition

Spectroscopy, a cornerstone of modern astronomy, involves dispersing light into its component wavelengths. By examining the resulting spectra, astronomers gain insights into the chemical composition of stars. Each element leaves a distinct imprint on the spectrum, enabling astronomers to identify the building blocks of stars and unravel the cosmic alchemy occurring within them.

Doppler Shifts and Stellar Motion

The Doppler effect, a shift in the observed wavelength of light due to the motion of a celestial object, provides a window into stellar dynamics. By analyzing these shifts, astronomers deduce the motion of stars—whether approaching or receding. This information contributes to our understanding of galactic rotation, the dynamics of binary star systems, and even the presence of exoplanets around distant stars.

Infrared Insights into Hidden Realms

In the cosmic quest for knowledge, astronomers turn to the infrared spectrum to unveil hidden realms. Infrared light penetrates cosmic dust clouds, revealing obscured regions where stars are born. The study's utilization of infrared observations, particularly through instruments like the Visible and Infrared Survey Telescope (VISTA), expanded the observational palette, allowing astronomers to peer into cosmic domains once veiled in mystery.

Cosmic Redshift and Expanding Universe

The cosmic redshift, observed as a shift of stellar spectra toward longer wavelengths, holds profound implications for our understanding of the universe. This phenomenon, attributed to the expansion of the cosmos, serves as evidence of an ever-expanding universe. The 10-year galaxy study, through meticulous spectroscopic analyses, contributed to our knowledge of the cosmic tapestry and its evolving dimensions.

Solar System-Sized Clouds: A New Type of Giant Star

In the celestial theater, a cosmic revelation emerged during the 10-year galaxy study, as astronomers uncovered a novel celestial performance starring a new type of giant star—dubbed "old smokers." At the heart of this discovery lies the intriguing presence of solar system-sized clouds, captivating in their enormity and unexpected origin.

A Stellar Surprise Unveiled

The enigmatic "old smokers," initially concealed within the bustling center of the Milky Way, presented a stellar surprise that defied conventional expectations. These elderly giants, previously unnoticed in their dormant phases, revealed a penchant for generating colossal solar system-sized clouds, adding an unexpected twist to the narrative of late-stage stellar evolution.

Quietude Followed by Cosmic Exhales

The narrative of "old smokers" unfolds in stages marked by quiet dormancy, during which these giant stars appear dim and red for extended periods. This tranquil phase is abruptly interrupted by episodes of cosmic exhales, where the stars puff out clouds of dust and gas on a scale comparable to our entire solar system. This alternating rhythm of quietude and exhalation sets these stellar performers apart in the cosmic ensemble.

Distinctive Location in the Galactic Center

"Old smokers" chose a specific stage for their cosmic performances—the innermost part of the Milky Way known as the Nuclear Disc. Concentrated in this metal-rich region, these giant stars exhibited their peculiar behavior, posing intriguing questions about the interplay between the stellar environment and the creation of solar system-sized clouds. The distinctive location of these giants added a layer of complexity to their cosmic narrative.

Convection Currents and Unraveling Mysteries

The mechanism behind the release of these monumental clouds remains a cosmic mystery, hinting at the role of convection currents and instabilities within the aging giants. The convoluted dance of matter on the surface of these stars triggers the release of vast columns of smoke. Unraveling the intricacies of these convection currents becomes a key to understanding the cosmic phenomenon of solar system-sized cloud formation.

Dust Condensation and the Galactic Symphony

The concentration of "old smokers" in the metal-rich Nuclear Disc raises questions about the process leading to the condensation of dust particles. Stellar giants, with their cool outer layers, offer favorable conditions for dust to condense. The resulting puffs of dense smoke, witnessed for the first time during the study, introduce a new crescendo in the galactic symphony, echoing the interconnectedness of stellar life cycles.

Red Giants with a Twist

The classification of "old smokers" as a new type of giant star marks a departure from the well-understood categories. Their distinctive behavior, characterized by extended dim and red phases followed by the spectacular release of solar system-sized clouds, sets them apart in the stellar panorama. This twist in the narrative of red giants challenges existing models, prompting astronomers to reassess the dynamics of stellar evolution.

Implications for Galactic Element Distribution

Beyond the stellar stage, the discovery of "old smokers" and their cosmic exhales carries broader implications. Matter ejected from these giant stars plays a crucial role in shaping the life cycle of elements, influencing the formation of the next generation of stars and planets. This unexpected source of solar system-sized clouds introduces a new variable in the spread of heavy elements within the galactic nucleus and metal-rich regions of other galaxies.

Characteristics of "Old Smokers": Unraveling the Enigmatic Giants

The discovery of "Old Smokers" in the heart of the Milky Way brought to light a new and enigmatic class of giant stars, challenging preconceived notions and introducing a cosmic twist to stellar evolution. As astronomers delved into their characteristics, a tapestry of unique features emerged, distinguishing these elderly giants within the celestial landscape.

Extended Dormancy and Dim Phases

"Old Smokers" exhibit prolonged periods of quiet dormancy, during which they appear dim and red. This extended phase of tranquility adds an element of mystery to their stellar behavior, as they quietly reside in the cosmic backdrop, often fading into near invisibility.

Abrupt Cosmic Exhales

The defining characteristic of "Old Smokers" lies in their unexpected and dramatic eruptions. These elderly giants, after years or decades of dormancy, undergo sudden outbursts, puffing out clouds of dust and gas on a colossal scale—comparable to the size of our entire solar system. This abrupt cosmic exhale sets them apart from conventional stellar behavior.

Distinctive Galactic Location

Concentrated in the innermost part of the Milky Way, known as the Nuclear Disc, "Old Smokers" choose a unique stage for their celestial performance. This location, rich in heavy elements, adds a layer of complexity to their characteristics, suggesting an intricate relationship between their behavior and the metal-rich environment.

Convection Currents and Surface Instabilities

The mechanism behind the release of solar system-sized clouds remains a cosmic puzzle. Convection currents and instabilities on the surface of these elderly giants are believed to play a crucial role. These convoluted dynamics trigger the extraordinary release of vast columns of smoke, contributing to the cosmic spectacle observed during their eruptions.

Matter Ejection and Element Distribution

"Old Smokers" contribute significantly to the life cycle of elements within the galaxy. The matter ejected during their eruptions plays a key role in seeding the next generation of stars and planets. This unexpected source of material challenges previous assumptions about the predominant role of certain types of stars in the distribution of heavy elements.

Heavy Concentration in the Nuclear Disc

The clustering of "Old Smokers" in the Nuclear Disc emphasizes their preference for a specific galactic locale. This concentration suggests a connection between the dynamics of these giants and the unique conditions found in this metal-rich region, where stars tend to be enriched in heavy elements.

Solar System-Sized Clouds as Cosmic Signatures

The solar system-sized clouds expelled by "Old Smokers" act as cosmic signatures of their eruptive episodes. These clouds, speculated to be composed of dust and gas, create a visual spectacle that challenges traditional notions of stellar behavior. The study of these clouds provides valuable insights into the late stages of stellar evolution and the cosmic processes shaping our galactic neighborhood.

Red Giants with Unconventional Behavior

The classification of "Old Smokers" as a new type of red giant star highlights their unconventional behavior. Their extended dim and red phases, coupled with the spectacular eruptions, mark them as outliers in the stellar population. This uniqueness prompts astronomers to reassess existing models and refine our understanding of the diverse paths stars can take in their evolutionary journey.

Theoretical Explanations for the Ejection of Dust and Gas by "Old Smokers"

The cosmic spectacle of "Old Smokers" puffing out solar system-sized clouds, observed during the 10-year galaxy study, has left astronomers intrigued and searching for theoretical explanations behind this enigmatic phenomenon. As these elderly giants undergo sudden and colossal eruptions, theories have been proposed to unravel the mechanisms triggering the ejection of dust and gas on such a grand scale.

Convection-Driven Dynamics

One prevailing theoretical explanation revolves around convection-driven dynamics on the surface of "Old Smokers." These red giants, characterized by convective cells in their outer layers, experience churning motions akin to giant bubbles. The convective currents and instabilities on the stellar surface could lead to the expulsion of material, forming solar system-sized clouds.

Surface Instabilities and Oscillations

The irregularities in the surface of aging stars like "Old Smokers" can give rise to surface instabilities and oscillations. These fluctuations may create localized disturbances, triggering the release of matter into space. The periodic nature of these oscillations could explain the observed pattern of extended dormancy followed by abrupt cosmic exhales.

Pulsation-Induced Mass Loss

Stellar pulsations, where a star undergoes periodic expansion and contraction, could play a role in the ejection of material. The pulsation-induced mass loss hypothesis suggests that during the contraction phase, material is expelled from the outer layers of the star, forming the observed solar system-sized clouds.

Interaction with Stellar Magnetic Fields

Stellar magnetic fields, inherent to many stars, could influence the ejection of matter from "Old Smokers." The interaction between convective currents and magnetic fields may create magnetic loops or structures that contribute to the expulsion of material. This magnetic influence adds a layer of complexity to the dynamics driving the eruptions.

Dust Condensation in Cool Outer Layers

The cool outer layers of red giants provide an environment conducive to the condensation of dust particles. The theoretical proposition involves the accumulation of dust in certain regions on the stellar surface. Subsequent eruptions release these accumulations, forming solar system-sized clouds composed of dust and gas.

Thermal Instabilities and Heating Processes

Thermal instabilities within the outer layers of "Old Smokers" may contribute to the ejection of material. Fluctuations in temperature, possibly triggered by internal processes or external influences, could lead to sudden heating events. These thermal instabilities may then propel matter into space, forming the observed clouds.

Influence of Stellar Companions

The presence of stellar companions or binary interactions could play a role in the ejection of material. Gravitational interactions between stars in close proximity might induce disruptions in the outer layers, contributing to the release of dust and gas. This theoretical avenue explores the influence of stellar companionship on the observed phenomena.

Location Matters: The Nuclear Stellar Disc and "Old Smokers"

In the cosmic drama unfolding at the heart of the Milky Way, the peculiar behavior of "Old Smokers" is intimately tied to their distinctive stage—the Nuclear Stellar Disc. This galactic locale, rich in heavy elements and nestled in the innermost part of our galaxy, emerges as a crucial backdrop influencing the characteristics and cosmic performances of these enigmatic elderly giants.

The Galactic Epicenter

The Nuclear Stellar Disc, positioned at the center of the Milky Way, stands as a gravitational epicenter teeming with stars. This densely packed region, known for its concentration of heavy elements, serves as a celestial theater where the cosmic narrative of "Old Smokers" unfolds. The proximity to the galactic center adds a layer of complexity to their behavior, distinguishing them from their counterparts in more distant galactic neighborhoods.

Metal-Rich Environs

The Nuclear Stellar Disc boasts an abundance of heavy elements, a cosmic treasure trove that sets the stage for the unique behavior of "Old Smokers." Stars in this region tend to be richer in metals, providing favorable conditions for dust particles to condense and form the solar system-sized clouds observed during the eruptions. The metal-rich environment contributes to the intricate cosmic symphony played out by these elderly giants.

Condensation Zones and Dust Dynamics

Within the Nuclear Stellar Disc, conditions are ripe for the condensation of dust particles. The cool outer layers of "Old Smokers" facilitate the accumulation of dust in certain regions on the stellar surface. As these giants undergo convective currents and surface instabilities, the interplay with the metal-rich environs triggers the release of vast columns of dust and gas, forming the characteristic solar system-sized clouds.

Influence on Element Distribution

The Nuclear Stellar Disc's role extends beyond the stellar stage, influencing the broader distribution of elements within the galaxy. Matter ejected during the eruptions of "Old Smokers" becomes a key player in the life cycle of elements, contributing to the formation of subsequent generations of stars and planets. The concentration of these events in the Nuclear Disc adds a unique chapter to the story of galactic evolution.

Concentration of Unique Cosmic Phenomena

The proximity to the galactic center positions the Nuclear Stellar Disc as a hotspot for unique cosmic phenomena. The concentration of "Old Smokers" in this region, exhibiting their peculiar behavior, hints at the interplay between galactic dynamics and the late stages of stellar evolution. The Nuclear Disc becomes a natural laboratory for studying the intricacies of these elderly giants and their impact on the galactic ecosystem.

Challenges to Existing Models

The prevalence of "Old Smokers" and their eruptions challenges existing models of stellar evolution, particularly those tailored to different galactic environments. The Nuclear Stellar Disc introduces a new variable that prompts astronomers to reassess their understanding of the factors influencing late-stage stellar dynamics. This recalibration is essential for comprehending the diverse paths stars can take within the dynamic landscape of the galactic center.

Significance of the Nuclear Disc in the Milky Way

In the celestial ballet of the Milky Way, the Nuclear Disc emerges as a pivotal player, influencing the galactic symphony and shaping the destiny of stars within its gravitational embrace. This compact and metal-rich region at the heart of our galaxy holds profound significance, impacting various aspects of galactic dynamics and stellar evolution.

Central Galactic Hub

The Nuclear Disc stands as the central hub of the Milky Way, anchoring the gravitational dynamics that govern the entire galaxy. Its concentrated mass plays a crucial role in shaping the orbits and trajectories of stars, contributing to the intricate dance of celestial bodies within this cosmic arena.

Rich in Heavy Elements

One defining feature of the Nuclear Disc is its abundance of heavy elements, also known as metals in astronomical terms. Compared to other regions of the galaxy, stars within the Nuclear Disc are enriched with elements beyond hydrogen and helium. This metal-rich environment sets the stage for unique astrophysical phenomena and influences the composition of stars born within its confines.

Formation of New Stellar Generations

The Nuclear Disc serves as a cosmic nursery, fostering the birth of new stellar generations. The enriched metal content provides the building blocks necessary for the formation of stars, planets, and other celestial bodies. The gravitational interactions within this region contribute to the dynamic process of stellar birth, shaping the evolving population of stars at the galactic center.

Influence on Stellar Dynamics

Stars in the Nuclear Disc experience a different gravitational environment compared to those in outer regions of the galaxy. This distinct gravitational influence affects their orbits, velocities, and interactions. The Nuclear Disc becomes a crucible where the dynamics of stellar clusters and individual stars are molded by the gravitational forces at play, leading to a unique stellar distribution.

Testing Ground for Galactic Models

The concentrated and complex nature of the Nuclear Disc provides a testing ground for galactic models and theories. Studying the behavior of stars in this central region offers insights into the broader dynamics of the Milky Way. It serves as a laboratory where astronomers can refine their understanding of galactic structures, stellar interactions, and the evolution of the galactic nucleus.

Interaction with Central Supermassive Black Hole

The Nuclear Disc is in close proximity to the supermassive black hole, Sagittarius A*, residing at the very center of the Milky Way. This proximity results in intricate gravitational interactions, influencing the behavior of stars and other celestial objects. Understanding this interaction contributes to our comprehension of the dynamics surrounding supermassive black holes and their impact on galactic evolution.

Unique Stellar Phenomena

The Nuclear Disc harbors unique stellar phenomena, such as the discovery of "Old Smokers." These elderly giants, exhibiting unexpected eruptions, add a layer of complexity to the galactic narrative. The peculiar behavior of stars within this region challenges existing models and prompted astronomers to explore the diverse paths stars can take in their evolutionary journey.

Crucial Element in Galactic Ecology

As a central player in the galactic ecology, the Nuclear Disc influences the distribution of matter, the evolution of stars, and the lifecycle of elements within the Milky Way. The concentration of heavy elements, coupled with its role as a stellar nursery, contributes to the intricate balance that sustains the galactic ecosystem.

Concentration of "Old Smokers" in the Innermost Galaxy: Galactic Enigma Unveiled

In the cosmic heart of the Milky Way, a captivating enigma unfolds as astronomers unveil the concentration of "Old Smokers" within the innermost reaches of our galaxy. This peculiar clustering of elderly giants, exhibiting their unique and unexpected behavior, adds a distinctive chapter to the narrative of stellar dynamics in the central regions of the Milky Way.

Dense Stellar Population

The innermost galaxy, characterized by its dense stellar population, sets the stage for the cosmic drama of "Old Smokers." Nestled within this gravitational cauldron, these elderly giants find themselves in a bustling environment, interacting with neighboring stars and influenced by the complex gravitational forces prevalent in the galactic nucleus.

Proximity to the Galactic Center

The concentration of "Old Smokers" near the Galactic Center enhances the intrigue surrounding their behavior. The gravitational influence exerted by the supermassive black hole, Sagittarius A*, contributes to the intricate dance of stars within this region. Proximity to this cosmic powerhouse adds a layer of complexity to the dynamics of "Old Smokers" and their eruptions.

Metal-Rich Environment

The innermost galaxy, including the Nuclear Disc, is known for its richness in heavy elements. This metal-rich environment provides favorable conditions for the formation of dust and gas, crucial components of the solar system-sized clouds expelled by "Old Smokers" during their eruptive episodes. The concentration of metals influences the cosmic performance of these elderly giants.

Close Stellar Interactions

The proximity of stars within the innermost galaxy increases the likelihood of close stellar interactions. Gravitational encounters between stars may play a role in triggering the eruptions of "Old Smokers." These interactions could induce disturbances in the outer layers of these elderly giants, leading to the sudden release of vast columns of dust and gas.

Formation within Galactic Structures

The clustering of "Old Smokers" within the innermost galaxy suggests a connection to specific galactic structures. Their preferential location hints at an intricate interplay of formation mechanisms, stellar dynamics, and environmental influences within the central regions. Unraveling the factors contributing to this concentration becomes essential in understanding the broader galactic ecology.

Influence of the Nuclear Disc

The proximity to the Nuclear Disc, a metal-rich and dynamic region, amplifies the mystery surrounding the concentration of "Old Smokers." The interplay between these elderly giants and the Nuclear Disc adds layers of complexity to their behavior, impacting the ejection of matter and the formation of solar system-sized clouds. The Nuclear Disc becomes a key player in shaping the narrative of "Old Smokers."

Challenges to Conventional Stellar Models

The concentration of "Old Smokers" challenges conventional stellar models, particularly those tailored to less dense galactic environments. Their unique behavior, coupled with their specific location, prompts astronomers to reassess existing theories and refine models to accommodate the intricacies observed within the innermost galaxy.

Implications for Galactic Evolution

The enigmatic concentration of "Old Smokers" holds implications for the broader evolution of the Milky Way. The matter ejected by these elderly giants contributes to the life cycle of elements, influencing the formation of subsequent generations of stars and planets. Understanding their concentration in the innermost galaxy provides valuable insights into the intricate processes shaping galactic ecosystems.

Implications for Element Distribution: The Cosmic Legacy of "Old Smokers"

The discovery of "Old Smokers" in the innermost galaxy not only captivates astronomers with their unique eruptions but also unveils profound implications for the distribution of elements across the cosmic landscape. These elderly giants, through their enigmatic behavior, play a pivotal role in shaping the abundance and dispersion of crucial building blocks essential for the formation of stars, planets, and the cosmic tapestry at large.

Elemental Enrichment in the Innermost Galaxy

The concentration of "Old Smokers" within the innermost galaxy signifies a localized source of elemental enrichment. The matter expelled during their eruptions, composed of dust and gas, becomes a cosmic reservoir of heavy elements. This localized enrichment contributes to a unique elemental fingerprint within the central regions, influencing the composition of subsequent stellar generations.

Seeding the Galactic Ecosystem

The solar system-sized clouds released by "Old Smokers" act as cosmic seeds, dispersing elements across the galactic ecosystem. As these clouds diffuse into the interstellar medium, they become the raw materials for the formation of new stars, planets, and other celestial bodies. The legacy of "Old Smokers" thus extends beyond their individual eruptions, leaving an indelible mark on the galactic canvas.

Galactic Recycling of Elements

The eruptions of "Old Smokers" initiate a cycle of galactic recycling, where the expelled material undergoes processes of condensation and incorporation into new stellar systems. This recycling mechanism ensures that the heavy elements generated by these elderly giants become integral to the formation of subsequent stellar and planetary bodies, perpetuating the cosmic cycle of element distribution.

Influence on Stellar Birth and Evolution

The enriched material from "Old Smokers" becomes a crucial factor in the birth and evolution of stars within the innermost galaxy. Stellar nurseries, influenced by the heavy elements released during eruptions, give rise to new stellar populations with distinctive compositions. The localized nature of this enrichment shapes the characteristics of stars emerging in the proximity of these cosmic phenomena.

Challenging Conventional Models

The discovery of "Old Smokers" challenges conventional models of element distribution within galaxies. Their concentration in specific regions introduces a dynamic element that influences the spatial variation of heavy elements. This challenges astronomers to refine existing models, accounting for the localized impact of these elderly giants on the galactic abundance of elements.

Insights into Galactic Evolution

The study of "Old Smokers" provides unique insights into the evolution of galaxies, particularly the innermost regions of the Milky Way. Their eruptions contribute to the ongoing processes shaping the galactic landscape, shedding light on the intricate interplay between stellar dynamics, element distribution, and the broader evolution of cosmic structures.

Linking Localized Phenomena to Galactic Dynamics

The localized nature of "Old Smokers" and their influence on element distribution offer a direct link between specific stellar phenomena and the broader dynamics of the innermost galaxy. Understanding this connection enhances our comprehension of how localized events contribute to the overall composition and evolution of galaxies, unraveling the intricacies of galactic ecology.

Potential Insights for Extragalactic Studies

The implications drawn from the study of "Old Smokers" extend beyond the Milky Way, offering potential insights for extragalactic studies. Similar phenomena in other galaxies could play comparable roles in shaping element distribution, opening avenues for comparative analyses and enriching our understanding of cosmic processes on a larger scale.

Role of Ejected Matter from "Old Smokers" in Stellar and Planetary Formation

The cosmic ballet of "Old Smokers" takes center stage not only for their intriguing eruptions but also for the pivotal role played by the matter they expel. As these elderly giants puff out solar system-sized clouds, the ejected material becomes a cosmic contributor, shaping the formation of stars and planets within the innermost galaxy.

Cosmic Alchemy

The matter ejected during the eruptions of "Old Smokers" embodies a celestial alchemy, carrying with it a diverse array of heavy elements. These elements, forged within the cores of stars and released into space through the aging process, become the building blocks for the next generation of celestial bodies, initiating a cosmic recycling of elements.

Stellar Nurseries Rejuvenated

The expelled material from "Old Smokers" acts as a rejuvenating agent for stellar nurseries within the innermost galaxy. As the clouds disperse into the interstellar medium, they seed regions where new stars are born. The enriched composition of these clouds influences the characteristics of the emerging stellar systems, shaping the fate of nascent stars and their planetary companions.

Formation of Protostars

The ejected matter provides the raw material for the formation of protostars—newborn stars in their earliest stages of development. As these protostars accrete material from their surroundings, the heavy elements introduced by "Old Smokers" become integral to the composition of the emerging stellar systems. The matter expelled by elderly giants thus initiates the transformative journey from cosmic dust to luminous stars.

Impact on Planetary Systems

Planetary systems, orbiting newly formed stars, benefit from the legacy of "Old Smokers." The heavy elements released during eruptions contribute to the protoplanetary disks—the cosmic cradles where planets take shape. This enriched environment influences the composition of planets, asteroids, and other celestial bodies within these burgeoning planetary systems.

Formation of Building Blocks

Within the protoplanetary disks, the ejected matter serves as the foundation for the formation of building blocks essential for planetary construction. Dust grains and gas particles, carrying the imprints of heavy elements from "Old Smokers," undergo processes of aggregation and coalescence. These building blocks become the precursors to planets, moons, and other celestial bodies in the making.

Influence on Planetary Composition

The composition of planets, influenced by the matter ejected by "Old Smokers," bears the imprint of the elderly giants' chemical legacy. The heavy elements introduced during the eruptions become integral components of planetary atmospheres, surfaces, and interiors. Understanding the chemical makeup of planets within the innermost galaxy requires consideration of the unique contributions from these cosmic phenomena.

Cosmic Life Cycle Continues

The matter expelled by "Old Smokers" becomes part of the ongoing cosmic life cycle. Elements released by aging stars contribute to the formation of subsequent generations of stars and planets, perpetuating the intricate dance of celestial bodies within the galactic ecosystem. The legacy of these elderly giants endures in the ongoing narrative of stellar and planetary evolution.

Insights into Galactic Ecology

Studying the interplay between "Old Smokers" and the formation of stars and planets provides valuable insights into the galactic ecology. The localized impact of these elderly giants on stellar nurseries, protostars, and planetary systems offers a microcosmic view of how specific stellar phenomena contribute to the broader dynamics of galactic evolution.

Revisiting Previous Models of Element Spread in Space: Insights from "Old Smokers"

The discovery of "Old Smokers" and their unexpected eruptions within the innermost galaxy has ushered in a new era of astronomical exploration, prompting a reevaluation of previous models governing the spread of elements in space. These enigmatic elderly giants, with their unique behavior, challenge established frameworks and offer valuable insights into the intricate processes shaping the distribution of elements within the cosmic expanse.

Unexpected Stellar Phenomena

The revelation of "Old Smokers," characterized by sudden eruptions and the release of solar system-sized clouds, stands in contrast to conventional expectations. The unexpected nature of these stellar phenomena challenges previous models that predominantly focused on well-understood events, urging astronomers to revisit and refine existing frameworks.

Localizing Element Ejections

The concentration of "Old Smokers" within specific regions of the innermost galaxy emphasizes the localized nature of element ejections. This challenges previous assumptions about the uniformity of element spread, highlighting the need to reconsider models that may have oversimplified the complexities of localized stellar dynamics and their impact on element distribution.

Impact on Elemental Abundance Patterns

The eruptions of "Old Smokers" introduce a unique pattern of elemental abundance within their vicinity. The matter expelled during these events contributes to the enrichment of nearby regions, influencing the composition of subsequent stellar and planetary systems. Revisiting previous models becomes essential to capture the nuances of how such localized phenomena shape the broader elemental landscape.

Integration of Gravitational Dynamics

The proximity of "Old Smokers" to the Galactic Center and their interactions with the Nuclear Disc underscore the intricate gravitational dynamics at play. Previous models may have overlooked the gravitational influences of specific galactic structures on element spread. Revisiting these models allows for a more comprehensive understanding of how gravitational interactions contribute to the observed patterns of element distribution.

Temporal Variability in Element Spread

The temporal variability exhibited by "Old Smokers," fading into invisibility for years before erupting, challenges assumptions about continuous and steady element spread.

Revisiting models to incorporate the temporal dynamics of stellar eruptions becomes crucial, recognizing that the cosmic stage is marked by episodic events that significantly impact element distribution over varying timescales.

Incorporating Unconventional Stellar Behaviors

"Old Smokers" represent a class of stars with unconventional behaviors not previously accounted for in standard models. Their sudden eruptions and solar system-sized clouds of dust and gas require a paradigm shift in our understanding of stellar evolution and element spread. Revisiting models provides an opportunity to incorporate these unconventional stellar behaviors into a more comprehensive framework.

Global Implications for Galactic Ecology

The localized impact of "Old Smokers" on element distribution in the innermost galaxy holds global implications for galactic ecology. Revisiting previous models extends beyond individual stellar phenomena, offering a broader perspective on how specific events contribute to the overall chemical evolution of galaxies. This reevaluation contributes to a more holistic understanding of galactic ecosystems.

Refining Predictions for Extragalactic Environments

Insights gained from revisiting models influenced by "Old Smokers" have the potential to refine predictions for element spread in extragalactic environments. Recognizing the diversity of stellar behaviors and their impact on element distribution enhances our ability to extrapolate findings beyond the Milky Way, contributing to a more accurate portrayal of cosmic chemical evolution on a larger scale.

Beyond Our Galaxy: Wider Significance Unveiled by "Old Smokers"

The discovery of "Old Smokers" and their unexpected behaviors within the Milky Way extends its reach beyond our galactic borders, unraveling wider significance that transcends the confines of our cosmic neighborhood. These elderly giants, with their enigmatic eruptions, offer profound insights into universal processes, enriching our understanding of stellar phenomena and element distribution on a grand cosmic scale.

Cosmic Analogues in Distant Galaxies

The study of "Old Smokers" prompts astronomers to explore the possibility of cosmic analogues in distant galaxies. Similar stellar phenomena, although unique in their manifestation, may exist beyond the Milky Way. The lessons learned from these elderly giants become valuable tools for interpreting observations and understanding the dynamics of stellar populations in diverse galactic environments.

Implications for Galactic Ecology Beyond the Local Group

The localized impact of "Old Smokers" on element distribution within the Milky Way serves as a template for broader galactic ecosystems. Extrapolating the lessons learned to galaxies beyond our Local Group offers a glimpse into the interconnectedness of stellar phenomena and their influence on the chemical evolution of cosmic structures spanning vast cosmic distances.

Diversity of Stellar Evolutionary Paths

The existence of "Old Smokers" challenges the notion of a universal stellar evolutionary path. Their unique behavior suggests a diversity in the life cycles of stars that extends beyond the familiar patterns observed within our galaxy. Exploring the wider significance involves considering how different galactic environments may host stars with unconventional behaviors, reshaping our understanding of stellar evolution on a universal scale.

Galactic Nuclei as Stellar Laboratories

The concentration of "Old Smokers" near the Galactic Center emphasizes the role of galactic nuclei as stellar laboratories. Beyond the Milky Way, investigating the stellar populations in the nuclei of other galaxies becomes essential. These cosmic laboratories may harbor similar elderly giants with unexpected behaviors, providing a rich canvas for studying the dynamics of stars near the gravitational epicenters of galaxies.

Contribution to Extragalactic Element Spread Models

Insights gained from studying "Old Smokers" contribute to the refinement of models predicting element spread in extragalactic environments. The localized impact of these elderly giants and their role in shaping the abundance of heavy elements offer a broader perspective on how such phenomena influence the chemical makeup of galaxies across the cosmic tapestry.

Understanding Universal Processes of Element Distribution

The eruptions of "Old Smokers" shed light on universal processes governing element distribution. Beyond the Milky Way, these insights become key components in deciphering the broader mechanisms shaping the chemical composition of galaxies. Understanding the universal nature of element spread enhances our ability to interpret observational data from diverse galactic landscapes.

Cosmic Connections Across the Cosmos

The discovery of "Old Smokers" establishes cosmic connections that reach across the cosmos. Lessons learned from their study serve as threads weaving a narrative that transcends individual galaxies. The interconnectedness of stellar phenomena and their influence on element distribution becomes a cosmic tapestry, linking galaxies in a shared story of stellar evolution and galactic ecology.

Perspectives for Future Observations and Surveys

The wider significance uncovered by "Old Smokers" shapes perspectives for future astronomical observations and surveys. Observing similar phenomena in galaxies beyond our own becomes a priority, offering opportunities to explore the diversity of stellar behaviors and their implications for galactic evolution. Future surveys may unveil new classes of stars with unexpected traits, expanding our cosmic repertoire.

Potential Impact on Understanding Other Galaxies: Lessons from "Old Smokers"

The discovery of "Old Smokers" within the Milky Way not only enriches our understanding of stellar phenomena within our galactic home but also holds the potential to significantly impact how we comprehend and explore other galaxies across the cosmic landscape. These elderly giants, with their unexpected eruptions, offer valuable lessons that can reshape the way we approach and interpret the complexities of stellar evolution and galactic dynamics beyond our immediate celestial neighborhood.

Diversifying Stellar Phenomena Catalogs

The existence of "Old Smokers" prompts astronomers to reconsider and expand catalogs of stellar phenomena in other galaxies. While previous classifications may have been anchored in well-understood events, the unique behavior of these elderly giants suggests that other galaxies could host a diverse array of stellar behaviors previously unaccounted for in observational databases.

Challenges to Conventional Models in Extragalactic Contexts

Extrapolating insights gained from "Old Smokers" challenges conventional models of stellar evolution and element distribution, especially in extragalactic environments. Observations in other galaxies may reveal similar phenomena that defy established paradigms, encouraging astronomers to refine models to accommodate the richness and

diversity of stellar behaviors witnessed across cosmic distances.

Refining Extragalactic Element Spread Predictions

The localized impact of "Old Smokers" on element distribution within the Milky Way offers a blueprint for refining predictions related to extragalactic element spread. As astronomers observe other galaxies, incorporating the lessons learned from these elderly giants enhances our ability to predict and interpret the distribution of heavy elements in diverse cosmic environments.

Exploring Galactic Nuclei in External Systems

The concentration of "Old Smokers" near the Galactic Center prompts a focused exploration of galactic nuclei in external galaxies. Observations of stellar populations in these nuclei may reveal analogous elderly giants with peculiar behaviors, providing insights into how galactic cores influence the dynamics and evolution of stars across a myriad of cosmic environments.

Comparative Galactic Ecology Studies
The discovery of "Old Smokers" sets the stage for comparative studies of galactic ecology beyond the Milky Way. Understanding how similar stellar phenomena contribute to the chemical evolution of galaxies with varying properties and histories becomes a key avenue for unraveling the universal principles governing the interconnectedness of stars and elements on a cosmic scale.

Identification of Unique Extragalactic Stellar Classes
Lessons from "Old Smokers" open the possibility of identifying unique stellar classes in other galaxies. The atypical behaviors exhibited by these elderly giants suggest that unconventional stellar traits may be more prevalent in the cosmic tapestry than previously assumed. Detecting and characterizing such unique stellar classes in external systems becomes a priority for future observational campaigns.

Impact on Extraterrestrial Life Studies

The localized nature of "Old Smokers" and their influence on the chemical makeup of galaxies could have implications for studies related to the potential for extraterrestrial life. Understanding the diversity of stellar behaviors in extragalactic environments informs the conditions under which planetary systems form, potentially influencing the habitability of exoplanets in other galaxies.

Guiding Future Extragalactic Surveys

The revelations from "Old Smokers" serve as a guide for planning and interpreting data from future extragalactic surveys. As astronomical instruments become more advanced, incorporating the understanding of localized phenomena and their impact on galactic dynamics will enhance the precision and depth of extragalactic observations, ushering in a new era of exploration.

Theoretical Connections to Metal-Rich Regions: Unveiling the Cosmic Tapestry

The discovery of "Old Smokers" and their peculiar behavior not only adds a unique chapter to stellar astrophysics within the Milky Way but also opens a theoretical gateway to exploring connections with metal-rich regions in galaxies at large. These elderly giants, with their unexpected eruptions, provide theoretical insights that may reshape our understanding of how heavy elements are distributed and influence stellar dynamics in regions abundant with metals across the cosmic tapestry.

Enrichment of Metal-Rich Galactic Regions

The concentration of "Old Smokers" in metal-rich regions, such as the Nuclear Disc of the Milky Way, sparks theoretical considerations about the role of heavy elements in shaping the behavior of stars. Theories may explore how the abundance of metals in these regions influences the formation, evolution, and ultimate fate of stars, including the peculiar phenomena exhibited by these elderly giants.

Interplay Between Metallicity and Stellar Eruptions

The theoretical connection between metallicity – the abundance of heavy elements – and the eruptions of "Old Smokers" becomes a focal point. Scientists may investigate how the interplay between metal-rich environments and stellar dynamics gives rise to the observed behaviors, unraveling the intricate relationship between the chemical composition of a region and the eruptive tendencies of its stellar inhabitants.

Dust Condensation in Metal-Rich Environments

Theoretical models may delve into the mechanisms behind the peculiar dust clouds expelled by "Old Smokers" within metal-rich surroundings. How do heavy elements facilitate the condensation of cosmic dust, leading to solar system-sized clouds? Theoretical explorations may seek to elucidate the underlying processes that give rise to these distinctive features in regions enriched with metals.

Formation of Metal-Rich Protostars

Theoretical considerations may extend to the formation of protostars within metal-rich environments influenced by "Old Smokers." How do heavy elements expelled by these elderly giants contribute to the birth and evolution of newborn stars in regions abundant with metals? Theoretical frameworks may aim to unveil the intricacies of protostellar formation influenced by the legacy of neighboring old stars.

Influence of Heavy Elements on Stellar Disc Stability

The localized eruptions of "Old Smokers" raise questions about the stability of stellar discs in metal-rich regions. Theoretical connections may explore how the ejection of matter, enriched with heavy elements, influences the stability and dynamics of surrounding stellar discs. This could offer insights into broader astrophysical processes shaping the evolution of stars within metal-rich galactic locales.

Metallicity as a Driver for Stellar Instabilities

Theoretical models may investigate whether metallicity plays a pivotal role in triggering instabilities within elderly stars, leading to the observed eruptions akin to "Old Smokers." Understanding how heavy elements contribute to the internal dynamics of stars could offer a broader perspective on the connection between metallicity and the manifestation of unique stellar behaviors.

Metal-Rich Regions as Stellar Laboratories

The theoretical exploration of metal-rich regions, influenced by the legacy of "Old Smokers," positions these locales as stellar laboratories. How do heavy elements impact the evolution of stars, and what role do these regions play in the broader context of galactic chemical enrichment? Theoretical connections may emphasize the significance of metal-rich environments as crucibles for understanding stellar phenomena.

Implications for Galactic Chemical Evolution Models

The theoretical implications extend to galactic chemical evolution models, where the role of "Old Smokers" in metal-rich regions may prompt revisions. How do these elderly giants contribute to the ebb and flow of heavy elements within galaxies? Theoretical frameworks may refine models to better capture the localized impact of stars on the chemical evolution of galactic environments enriched with metals.

Future Directions in Astrophysics: Navigating the Cosmic Frontier

The discovery of "Old Smokers" and the ongoing advancements in astrophysics propel the field toward new horizons, beckoning scientists to chart future directions that promise groundbreaking insights into the mysteries of the universe. As technology and methodologies evolve, the following areas represent key avenues for exploration and discovery in the coming era of astrophysics:

Advanced Telescopic Technologies

Future astrophysics will witness the deployment of cutting-edge telescopic technologies with enhanced sensitivity, resolution, and coverage across the electromagnetic spectrum. From space-based observatories to ground-based mega-telescopes, these instruments will revolutionize our ability to observe distant galaxies, exoplanets, and transient celestial events with unprecedented clarity.

Multimessenger Astronomy

The integration of gravitational wave detectors, neutrino observatories, and traditional telescopes marks a new era in multimessenger astronomy. Future endeavors will focus on refining this interdisciplinary approach, enabling the simultaneous observation of cosmic phenomena using multiple messengers. This promises a deeper understanding of events such as mergers of compact objects, supernovae, and other cataclysmic occurrences.

Precision Cosmology and Dark Energy Studies

Advancements in precision cosmology will delve into the nature of dark energy, unraveling its enigmatic properties and influence on the expansion of the universe. Future missions and experiments aim to refine measurements of cosmic microwave background radiation, cosmic large-scale structures, and supernovae, providing crucial data for unraveling the cosmic tapestry.

Exoplanet Characterization and Habitability Studies

The focus on exoplanet characterization will intensify, with upcoming space missions designed to scrutinize atmospheres and surfaces of distant worlds. The search for biosignatures and habitable environments will guide future directions, probing the potential for life beyond our solar system and redefining our understanding of planetary systems in the Milky Way and beyond.

Quantum Astrophysics and Information Theory

The convergence of quantum technologies with astrophysics will open new avenues for exploration. Quantum sensors, communication, and computing may revolutionize observational capabilities, allowing scientists to tackle complex computational challenges and explore quantum aspects of celestial phenomena, pushing the boundaries of information theory in astrophysical contexts.

Astroinformatics and Big Data Analytics

The exponential growth of observational data necessitates the development of sophisticated astroinformatics tools and big data analytics. Future directions will focus on artificial intelligence, machine learning, and data mining techniques to extract meaningful insights from vast datasets, facilitating discoveries in transient events, variable stars, and other dynamic astrophysical phenomena.

Astrobiology and the Search for Extraterrestrial Life

The search for extraterrestrial life will continue to be a pivotal focus in future astrophysics. Upcoming missions to Mars, the study of icy moons, and advancements in astrobiology will guide our understanding of habitability in our solar system and beyond. Future directions may involve refining biosignature detection techniques and exploring the potential for life in diverse cosmic environments.

Interstellar Travel and Robotic Exploration

As humanity looks beyond the confines of our solar system, future directions in astrophysics may involve the development of propulsion technologies for interstellar travel. Robotic exploration missions to nearby exoplanetary systems, propelled by novel propulsion methods, may become a reality, opening new frontiers for scientific exploration and expanding our cosmic perspective.

Astrophysical Gravitational Laboratories

The study of extreme gravitational environments, such as black holes and neutron stars, will advance with the development of gravitational laboratories in space. Future missions, like the Laser Interferometer Space Antenna (LISA), will enable the detection of gravitational waves from sources not observable on Earth, providing insights into the fundamental nature of spacetime.

International Collaboration and Space Agencies

Future directions in astrophysics will be characterized by increased international collaboration, with space agencies worldwide pooling resources and expertise. Joint missions, shared observatories, and collaborative research endeavors will foster a global approach to addressing fundamental questions about the cosmos, transcending individual nations and uniting the global scientific community.

Unanswered Questions and Frontiers for Further Astrophysical Research

The exploration of the cosmos has brought forth a tapestry of discoveries, yet numerous unanswered questions persist, beckoning astronomers and researchers to delve deeper into the mysteries of the universe. As we peer into the cosmic unknown, several key areas emerge as frontiers for further astrophysical research:

Nature of Dark Matter and Dark Energy

The elusive nature of dark matter and dark energy continues to confound astrophysicists. Unraveling the properties of these mysterious components, which constitute the majority of the universe's mass-energy content, remains a paramount challenge. Future research will aim to discern the fundamental nature of dark matter and dark energy and their roles in shaping the cosmos.

Origin of Cosmic Magnetism

The origin and evolution of cosmic magnetic fields pose intriguing puzzles. Researchers seek to understand how these magnetic fields influence the formation and dynamics of galaxies, stars, and even the cosmic web. Unraveling the enigma of cosmic magnetism is a frontier where further observational and theoretical studies are crucial.

Quantum Nature of Black Holes

While classical physics describes black holes well, the interface between quantum mechanics and gravity within the vicinity of black holes remains enigmatic. Investigating the quantum nature of black holes, exploring the information paradox, and understanding the dynamics near the event horizon are fertile grounds for future research in theoretical astrophysics.

Unified Theory of Fundamental Forces

The quest for a unified theory that reconciles general relativity and quantum mechanics continues to be a central challenge. Researchers aspire to formulate a framework that seamlessly integrates the fundamental forces of the universe, providing a comprehensive understanding of cosmic phenomena from the smallest to the largest scales.

Detection and Characterization of Exoplanets

While thousands of exoplanets have been discovered, understanding their atmospheres and surfaces remains a frontier. Further research will focus on refining techniques for exoplanet characterization, searching for signs of habitability, and probing the potential presence of biosignatures. Unraveling the diversity of exoplanetary systems holds key insights into the prevalence of life beyond Earth.

Origin of Fast Radio Bursts (FRBs)

Fast Radio Bursts, cosmic radio signals of unknown origin and extreme intensity, pose a tantalizing mystery. Identifying the sources and mechanisms behind FRBs remains an active area of investigation. Future research will strive to pinpoint the origins of these enigmatic signals and elucidate the astrophysical processes responsible for their sudden and intense bursts.

Understanding Dark Fluid and Cosmic Acceleration

The accelerated expansion of the universe raises questions about the nature of dark energy. Exploring alternative theories, such as dark fluid models, to explain cosmic acceleration is an evolving frontier. Researchers aim to refine our understanding of the interplay between dark energy, cosmic expansion, and the ultimate fate of the universe.

Cosmic Microwave Background (CMB) Anomalies

Anomalies in the cosmic microwave background radiation challenge our current cosmological models. Investigating the origins of these anomalies, such as the Cold Spot and hemispheric asymmetries, will guide future research in precision cosmology. Addressing these discrepancies may unveil new insights into the early universe and its evolution.

Nature of High-Energy Astrophysical Phenomena

High-energy astrophysical phenomena, including gamma-ray bursts, cosmic rays, and ultra-high-energy cosmic particles, harbor unresolved questions about their origins and acceleration mechanisms. Future research will delve into the extreme realms of the universe, utilizing advanced observatories and detectors to unveil the secrets behind these energetic cosmic phenomena.

Evolution of Stellar Populations in Diverse Environments

Understanding the evolution of stars in diverse galactic environments remains a complex puzzle. Investigating how factors like metallicity, stellar density, and gravitational interactions shape stellar populations will drive further research. Exploring the life cycles of stars in various cosmic contexts holds the key to comprehending galactic dynamics and chemical evolution.

Technological Advancements for Deeper Astrophysical Insights

The pursuit of deeper insights into the mysteries of the cosmos requires continuous advancements in technology. As astronomers strive to unravel the complexities of the universe, several key technological developments are crucial for pushing the boundaries of astrophysical knowledge:

Next-Generation Telescopes

The construction of larger, more powerful telescopes with advanced optics and adaptive optics systems is essential. Telescopes like the James Webb Space Telescope (JWST) and extremely large ground-based telescopes promise enhanced sensitivity and resolution, enabling the observation of fainter celestial objects and finer details within distant galaxies.

High-Resolution Imaging Techniques

Advancements in high-resolution imaging techniques, such as interferometry and novel imaging algorithms, are vital. These technologies enable astronomers to capture detailed images of astronomical objects, including exoplanets, galactic centers, and complex structures within the universe.

Quantum Sensors and Computing

The integration of quantum sensors and quantum computing can revolutionize data acquisition and analysis. Quantum sensors offer unprecedented sensitivity for detecting faint signals, while quantum computing accelerates complex calculations required for simulating astrophysical phenomena and processing vast datasets.

Advanced Spectroscopy Instruments

Upgrading spectroscopy instruments with higher spectral resolution and broader wavelength coverage enhances the ability to study celestial objects in detail. Advanced spectrographs, such as those with integral field spectroscopy capabilities, provide valuable insights into the chemical composition, temperature, and motion of astronomical sources.

Multimessenger Astronomy Facilities

Expanding facilities for multimessenger astronomy, including gravitational wave detectors, neutrino observatories, and cosmic-ray detectors, amplifies our ability to study cosmic events from multiple perspectives. Coordinated observations across different messengers offer a comprehensive understanding of phenomena like neutron star mergers and supernovae.

Space-Based Gravitational Wave Detectors

Developing space-based gravitational wave detectors, like the Laser Interferometer Space Antenna (LISA), extends the frequency range accessible for gravitational wave observations. This technology enables the detection of low-frequency gravitational waves from massive black hole mergers and other sources not detectable by ground-based detectors.

Astroinformatics and Big Data Tools

The evolution of astroinformatics tools and big data analytics is crucial for handling the massive datasets generated by modern observatories. Machine learning algorithms, artificial intelligence, and advanced data mining techniques aid in extracting meaningful patterns from complex astrophysical datasets.

High-Performance Computing for Simulations

High-performance computing facilities are essential for running complex simulations of astrophysical processes. Advancements in computing power enable researchers to model intricate scenarios, such as galaxy formation, stellar evolution, and cosmological simulations, with greater accuracy and detail.

Space-Based Exoplanet Observatories

Dedicated space-based observatories for exoplanet research, equipped with advanced instruments for spectroscopy and direct imaging, are vital. These observatories can characterize exoplanet atmospheres, study their compositions, and identify potential signs of habitability or biosignatures.

Advancements in Cryogenic Technologies

Cryogenic technologies play a pivotal role in space-based observations, especially for instruments like infrared telescopes. Innovations in cryocoolers and cooling systems ensure that detectors and instruments can operate at extremely low temperatures, enhancing their sensitivity to infrared radiation.

Next-Generation Space Exploration Technologies

Developments in propulsion technologies, robotics, and autonomous systems are essential for the next generation of space exploration. These technologies enable extended missions, robotic exploration of distant celestial bodies, and the deployment of observatories in strategic locations, such as Lagrange points.

Advances in Spaceborne Interferometry

Spaceborne interferometry missions can significantly improve angular resolution and enable detailed observations of celestial objects. Innovations in deploying interferometric capabilities in space contribute to a deeper understanding of complex structures within galaxies and the dynamics of binary star systems.

Enhancements in Detector Technologies

Ongoing advancements in detector technologies, including semiconductor detectors and photon-counting detectors, improve sensitivity and speed for various wavelengths. These detectors are crucial for capturing faint signals, conducting precise measurements, and exploring the full electromagnetic spectrum.

Robotic Telescope Networks

Robotic telescope networks equipped with autonomous scheduling and adaptive optics enhance the efficiency of observational campaigns. These networks enable continuous monitoring of celestial events, rapid follow-up observations, and collaborative efforts across global observatories.

Advanced Materials for Space Instruments

The development of lightweight and durable materials for constructing space instruments is essential for future space missions. Advanced materials contribute to the design of robust and efficient instruments, ensuring the longevity and reliability of space-based observatories.

Conclusion: Navigating the Cosmic Odyssey

In the quest to understand the vast expanse of the cosmos, the journey of astrophysics unfolds as a perpetual odyssey, marked by discovery, inquiry, and the relentless pursuit of knowledge. From the recent revelation of "Old Smokers" to the ongoing exploration of gravitational waves and distant exoplanets, the tapestry of our understanding continues to weave itself, revealing threads of cosmic wonders and enigmas.

The 10-year galaxy study, led by the University of Hertfordshire, exemplifies the collaborative spirit of global scientific endeavors. Across borders, cultures, and continents, astronomers joined forces to scrutinize nearly a billion stars, leading to the discovery of hidden giants, erupting protostars, and the emergence of a new stellar archetype – the intriguing "Old Smokers." This astronomical odyssey not only broadens our comprehension of the Milky Way but also sparks theoretical reflections on the distribution of elements across cosmic realms.

As we stand at the cusp of the future, technological advancements beckon the next phase of astrophysical exploration. From the deployment of next-generation telescopes to the integration of quantum sensors, these innovations promise deeper insights into the fundamental nature of the universe. The uncharted territories of dark matter, quantum black holes, and the mysteries of cosmic acceleration beckon researchers to push the boundaries of our current understanding.

Yet, amidst the unanswered questions and technological frontiers lies the beauty of the cosmic journey – an ongoing narrative that transcends generations. The conclusion of one discovery sparks the inception of another inquiry, propelling humanity further into the cosmic odyssey. The interconnectedness of theoretical frameworks, observational prowess, and technological ingenuity forms the compass guiding us through the cosmic unknown.

In this conclusion, we find a call to action – a call for continued collaboration, relentless curiosity, and an unwavering commitment to advancing the frontiers of astrophysics. The cosmic odyssey is far from over; it is an ever-evolving narrative that invites scientists, scholars, and enthusiasts to contribute to the unfolding chapters of the cosmic story.

As we peer into the cosmic abyss, let the spirit of exploration guide us, and may the mysteries yet to be unraveled inspire the next generation of cosmic adventurers. The odyssey continues, and the cosmic tapestry awaits its next revelation.

Recap of Key Discoveries: Unveiling the Cosmic Secrets

The 10-year galaxy study, led by the University of Hertfordshire and an international team of astronomers, has unraveled a tapestry of celestial marvels, pushing the boundaries of our understanding of the Milky Way. As we reflect on this cosmic odyssey, let's recap the key discoveries that have illuminated the cosmic stage:

Hidden Giants – "Old Smokers"

The revelation of a new class of giant stars, endearingly nicknamed "Old Smokers," took center stage. These elderly giants, residing near the heart of the Milky Way, exhibit a peculiar behavior of remaining dormant for years before abruptly releasing solar system-sized clouds of dust and gas. This unexpected stellar phenomenon challenges existing models and prompts a reevaluation of the life cycles of aging stars.

Erupting Protostars

The survey unveiled 32 erupting protostars, newborn celestial entities undergoing extreme outbursts over months, years, or decades. These dynamic events, intricately linked to the formation of new solar systems, provide a unique opportunity to observe the birth pangs of stars and the challenges faced by emerging planets.

Multinational Collaboration

The collaborative effort spanning the United Kingdom, Chile, South Korea, Brazil, Germany, and Italy exemplifies the power of multinational collaboration in modern astrophysics. The fusion of diverse expertise, observational resources, and technological capabilities has enriched the exploration of our galaxy and beyond.

Revolutionizing Observation with Infrared Light

The study showcased the pivotal role of infrared light in unveiling celestial secrets. By harnessing the capabilities of the Visible and Infrared Survey Telescope (VISTA) and the European Southern Observatory's Very Large Telescope, astronomers pierced through cosmic dust and gas, revealing hidden stars and providing unprecedented insights into the universe's hidden realms.

Unprecedented Long-Term Monitoring

The researchers conducted unprecedented long-term monitoring of nearly a billion stars, capturing the nuances of stellar behavior over extended periods. This observational dedication allowed the identification of subtle changes in brightness, leading to the discovery of both erupting protostars and the mysterious "Old Smokers."

Spectroscopic Analysis with Precision

The team employed advanced spectroscopic techniques, including the use of the Very Large Telescope, to conduct detailed analyses of individual stars. Spectroscopic data provided crucial information about the composition, temperature, and dynamics of the observed celestial objects, offering a deeper understanding of their nature.

Concentration in the Nuclear Disc

The concentration of "Old Smokers" in the innermost part of the Milky Way, known as the Nuclear Disc, added an intriguing layer to the discovery. The prevalence of these enigmatic stars in a region rich in heavy elements poses questions about the mechanisms leading to the ejection of dust and gas, challenging existing models of element distribution.

Implications for Element Distribution

The discovery of a new type of giant star with the ability to eject matter has profound implications for the distribution of heavy elements in the cosmos. The traditional understanding, centered around Mira variables, is now complemented by the unexpected contributions of "Old Smokers," potentially reshaping our comprehension of element spread in other galaxies.

Future Directions and Wider Significance

The discoveries open avenues for future astrophysical exploration, from refining our understanding of stellar evolution to delving into the implications for other galaxies. The "Old Smokers" may hold the key to unlocking broader insights into the life cycle of elements, the formation of stars and planets, and the interconnectedness of cosmic processes.

Reflections on the Evolving Field of Astrophysics: Navigating Cosmic Frontiers

The field of astrophysics stands as a testament to humanity's enduring quest to comprehend the cosmos. As we reflect on the recent discoveries from the 10-year galaxy study and the broader trajectory of astrophysical exploration, several reflections emerge, guiding our understanding of the evolving cosmic landscape:

The Pervasiveness of Cosmic Surprises

The universe, with its boundless expanse, continues to surprise us with celestial phenomena that defy our expectations. The discovery of "Old Smokers" highlights the ever-present potential for the unexpected, challenging astronomers to adapt their models and theories to accommodate the richness of cosmic diversity.

Multidisciplinary Collaborations as Catalysts

The collaborative spirit exhibited in the 10-year galaxy study underscores the importance of multidisciplinary approaches in modern astrophysics. As the field becomes more complex, collaboration across borders, disciplines, and technological domains emerges as a catalyst for breakthroughs. Global partnerships amplify the collective intelligence needed to decipher the intricacies of the universe.

Technological Mastery Unveils Cosmic Mysteries

Technological advancements have become the guiding torchbearers in astrophysical exploration. From the revolutionary insights provided by infrared light observations to the precision of spectroscopic analyses with advanced telescopes, technology has elevated our ability to observe, measure, and interpret the cosmos. The continuous mastery of cutting-edge tools is pivotal for navigating deeper into cosmic frontiers.

From Singular Observations to Holistic Understanding

The era of singular observations has evolved into a pursuit of holistic understanding. Long-term monitoring, spectroscopic analyses, and multimessenger astronomy converge to provide a comprehensive narrative of cosmic events. The integration of various observational techniques allows astronomers to piece together the complete story, from the birth of stars to the dynamics of galactic centers.

Unraveling Mysteries with Every Discovery

Each discovery in astrophysics acts as a lantern illuminating the shadows of cosmic mysteries. The identification of "Old Smokers" prompts a reconsideration of our assumptions about aging stars, element distribution, and the interconnectedness of cosmic processes. With every revelation, new questions emerge, propelling us into an endless cycle of inquiry.

Balancing Theoretical Frameworks and Observational Realities

The evolving field demands a delicate balance between theoretical frameworks and observational realities. As our understanding of the universe deepens, theoretical models must adapt to accommodate the nuances revealed by precise observations. This dynamic interplay ensures that our cosmic narratives remain rooted in both theoretical foundations and empirical evidence.

Education as a Cosmic Gateway

The field of astrophysics not only unravels the secrets of the universe but serves as a cosmic gateway for education and inspiration. Each discovery invites the next generation of scientists, scholars, and enthusiasts to partake in the exploration of the unknown. Astrophysics, with its captivating narratives, fosters a sense of wonder and curiosity that transcends disciplinary boundaries.

Humility in the Face of Cosmic Grandeur

The grandeur of the cosmos humbles humanity in the face of its vastness and complexity. As we gaze into the depths of space, there is an inherent acknowledgment of the limits of our current understanding. This humility fuels the perpetual drive to explore, learn, and expand the frontiers of astrophysics.

Dear Readers and Cosmic Explorers,

Thank you for joining the cosmic odyssey in "Cosmic Odyssey: Unveiling Secrets of the Milky Way" by Emmons Rivas. Your journey through the mysteries of the universe is not just a solitary venture—it's a shared exploration that echoes through the cosmos.

As you navigate the pages of this book, we invite you to share your thoughts, reflections, and impressions by leaving a review on Amazon. Your reviews contribute not only to the vibrant community of cosmic enthusiasts but also to the broader understanding of the cosmos.

Why Your Review Matters

Community Building: Connect with fellow readers who share your passion for astrophysics and the wonders of the Milky Way. Your insights can spark conversations and foster a sense of shared exploration.

Author Feedback: Emmons Rivas values your feedback. Your reviews provide valuable insights into what resonated with you, helping the author refine their storytelling and engage with readers on a deeper level.

Guide for Others: Your review can be a guiding light for others contemplating their cosmic journey. Share your experience, highlight key takeaways, and help potential readers decide if this book aligns with their interests.

How to Leave a Review

1. **Visit Amazon:** Go to the "Cosmic Odyssey" product page on Amazon.
2. **Scroll Down:** Navigate to the Customer Reviews section.

3. **Write Your Review:** Click on "Write a customer review" and share your thoughts.
4. **Rate the Book:** Assign a star rating based on your overall experience.

Your Cosmic Voice Matters!

Whether you're an amateur stargazer, a seasoned astronomer, or someone captivated by the mysteries of the cosmos, your perspective is invaluable. Let your cosmic voice resonate by leaving a review and contributing to the ongoing cosmic dialogue.

Thank you for being a part of this cosmic exploration!

Warmest Regards,
Emmons Rivas.